Maggie and the Magnifying Glass
Signs of Spring

ILLUSTRATED BY
MONTANA DEBOR

WRITTEN BY
MELISSA CAMPBELL

This book is dedicated to my mom and my mother-in-law. These women were both strong influences on me and have shaped the person I am today.

It's a fresh, cool morning,
she can hear the birds sing.

Maggie looks out her window
at the first signs of spring!

A butterfly flutters and lands so still,
soaking up the sun on the window sill.

Delighted by its colors, shapes, and more,
she grabs her magnifying glass
and heads out the door.

Maggie runs outside and smells the green grass.
She is off to explore with her magnifying glass!

Spring is a beautiful sight to see,
with ladybugs, flowers, and leaves on the trees.

Maggie spots a fern
growing out of the ground,

its stems and leaves
twisting all around.

She crouches down to take a closer look,

and sees the fern curled up like a hook.

Spinning and twirling through the tall grass,
she is off to explore with her magnifying glass!

TWIST! TWIRL!

FLUTTER! FLUTTER!

Maggie stops at an iris that's bright as a gem,
and sees a water drop resting on the stem.

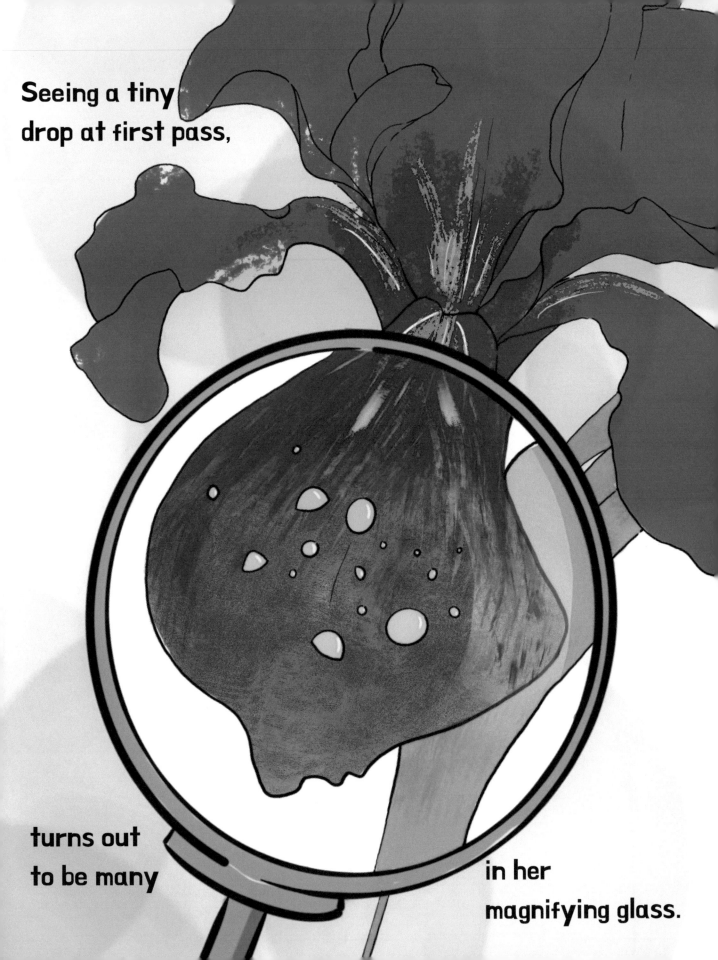

Seeing a tiny
drop at first pass,

turns out
to be many

in her
magnifying glass.

As she explores the colors purple and white ,
the iris petals feel so soft and light.

The gentle wind blows and the flowers swish.
Maggie picks a dandelion and makes a **BIG** wish.

As the delicate seeds float to the grass,

Maggie's off to explore
with her magnifying glass!

She watches a dragonfly

twist and flip,

as it stops to rest on a bright red tulip.

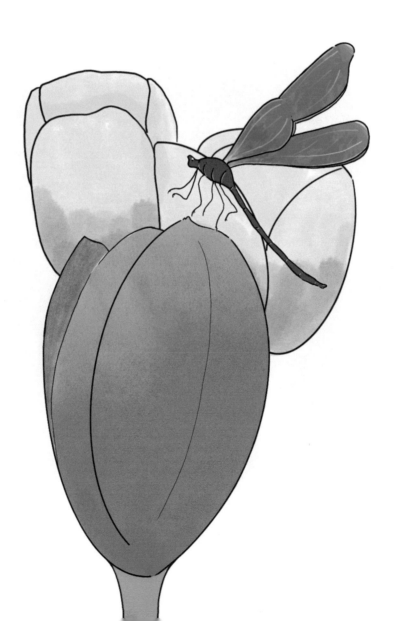

A yellow butterfly floats gracefully in,
and lands beside her dragonfly friend.

Through the flowers they dance and play.

Maggie smiles and winks as they fly away.

She continues her journey to the top of a hill,
amazed by the vibrance of daffodils.

Maggie spies a grasshopper resting mellow,
on a flower petal, bright and yellow.

Curious to see, she decides to stop,
but the grasshopper bounces away with a HOP!

Maggie laughs and runs amongst the tall grass.
She is off to explore with her magnifying glass!

Upon a grass blade,
she sees something squirm.

Up close she observes
a tiny inch worm.

He **wiggles** and **shuffles** along the way,
munching and nibbling his food for the day.

Looking up at the sky there is beauty to see.

Maggie spots white flowers up in a tree!

A curious flower,
this dogwood bloom.

She pulls the limb closer
for a magnified zoom.

As heavy rain drops
begin to fall hard,
Maggie heads inside from
her backyard.

Off to her room, she flops on her bed.
Memories of her adventure fill her head.

She reflects on all the things she found,
while exploring signs of spring all around.

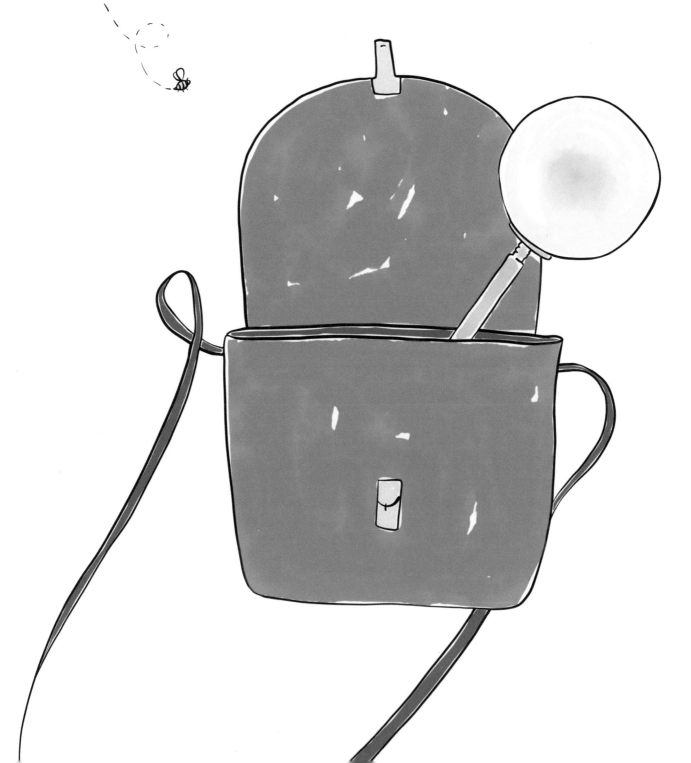

Made in the USA
Columbia, SC
16 August 2024

40100144R00020